Drone Pilots

Liana Robinson

Series Editor **Casey Malarcher**

Level 2 - 1

Drone Pilots

Liana Robinson

© 2018 Seed Learning, Inc.

Series Editor: Casey Malarcher
Acquisitions Editor: Anne Taylor
Copy Editor: Kelly Daniels
Cover/Interior Design: Highline Studio

ISBN: 978-1-943980-38-3

10 9 8 7 6 5 4 3 2 1
22 21 20 19 18

Photo Credits
All photos are © Shutterstock, Inc.

Contents

Drones

Do you have a drone? Have you ever flown one before? They are a lot of fun.

◀ Flying a drone for fun

Children can fly drones.

People can fly drones with the use of a remote control. Drones can also fly themselves using the right computer programs.

◀ A drone with a camera

Drones can take photos or videos. They can also deliver packages.

◀ A drone carrying a package

The Job of a Drone Pilot

Drone pilots are people who fly drones. They fly drones in cities and rural areas. Many drone pilots spend a lot of time flying their drones outside. They travel to exciting and beautiful places.

Flying a drone in ▶
a park

Flying a drone by a river

If spending time outside does not interest you, there are a lot of other drone pilots who rarely leave their offices.

Looking at pictures ▶
taken by a drone

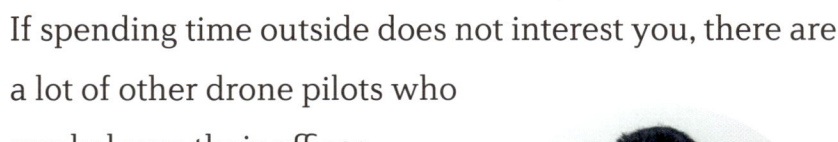

◀ Operating a drone
from an office

Drones That Help People

In cities, drone pilots work for many different companies including builders and city governments. Builders get drone pilots to map the land they want to build on. They also use drones to monitor big projects such as the building of skyscrapers.

◄ What a drone sees from high above a city

A drone by a ►
building project

A skyscraper

More and more city governments are beginning to work with drone pilots. These pilots monitor floods and other natural disasters.

A drone searching in a national park

◀ A building on fire

Police drones ▶
in action

Drone pilots are also beginning to work with firefighters, police forces, and lifeguards. During a fire, drones can check for danger before firefighters go in. In the future, drone pilots may give out traffic tickets and do other police work.

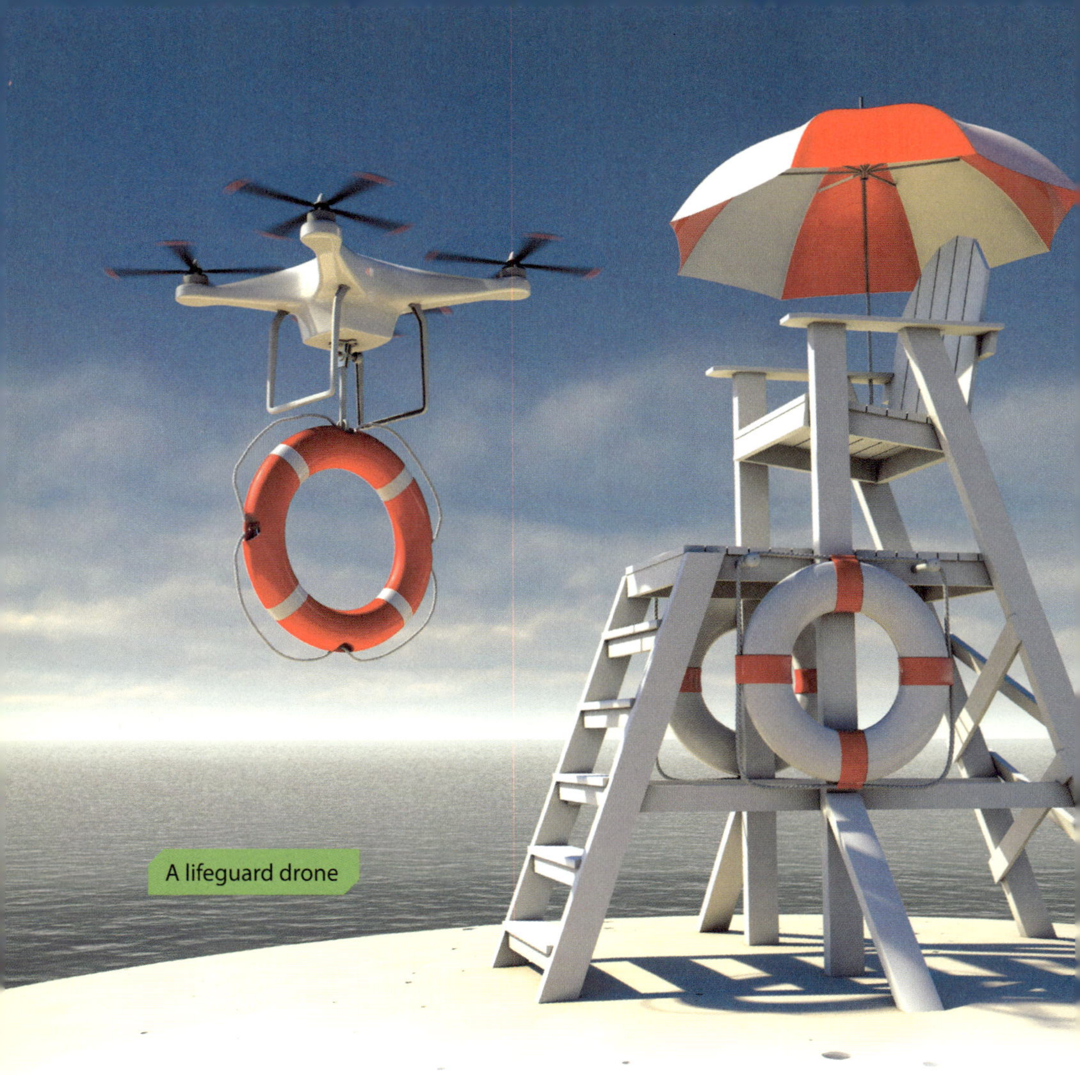

A lifeguard drone

Lifeguards use drone pilots in difficult rescue situations. There is a drone that can fly above the water and help a drowning person.

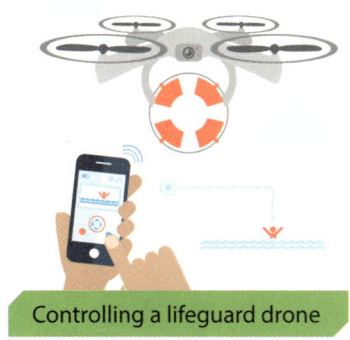

Controlling a lifeguard drone

Drones That Explore and Do Research

Drone pilots work in rural areas as well. They can work for farmers, animal and nature projects, search and rescue teams, and explorers.

Using a drone camera to view land

High Yield
Medium Yield
Low Yield

Drone pilots can help farmers in many ways. Drones can be used to monitor the soil, plant seeds, and tell farmers when the crops are ready. Drones can also be used to keep track of animals and make sure they are healthy.

Watching animals with a drone ▶

A farmer with a drone

The view of nature from a drone camera

Animal and nature projects can work with drone pilots to monitor wild animals and forests. Drone pilots can help protect animals and natural places. They can call the police as a crime is happening. Drone pilots can stop people from killing animals or cutting down trees on protected lands.

A drone with a first aid kit

Searching with a drone ▶

Drone pilots are also helpful to search and rescue teams. Drones can help find people who are lost. They can fly into dangerous places. They can even deliver ropes, water, food, and medicine to people in danger.

The army also has many uses for drone pilots. Drones can take aerial photographs and videos before soldiers go into dangerous places. There are also larger drones that carry weapons.

A military drone pilot ▶

Soldiers working with a drone

A drone's view of a waterfall

A drone explorer is a person whose job is to use drones to travel to unknown places. People work with drone explorers to make maps, extreme films, and more. Drones can fly into unknown places and use computer programs to make maps.

A map made with a ▶
drone's help

18

A drone's view from inside a cave

A drone's view of a volcano

Drones are often used in extreme films. They can fly into interesting and hard-to-reach places. For example, drones can fly over volcanoes and through small spaces such as caves.

An archaeologist

Drone explorers have even worked with archaeologists. They fly drones with special cameras over the earth to figure out where to dig. Drone explores travel all over the world for their jobs.

How to Become a Drone Pilot

Anyone can buy a drone and learn to fly it. It is a fun hobby. Some middle schools and high schools even have basic drone courses or drone clubs. However, working as a drone pilot as a job is more difficult. People need to complete courses and know all the drone laws. There are, for example, places where it is against the law to fly drones.

Flying a drone as a hobby

Universities and private companies offer drone pilot courses. But these courses are not like classes for flying planes. Drone pilots do not need to take the same courses as aircraft pilots.

A pilot ▶

Doing drone research

Some drone courses only teach people who wish to fly drones with remote controls. Other courses teach computer skills for people who wish to pilot drones that fly themselves. There are also coding courses for people who wish to program drones that fly themselves.

Editing photos and videos on a computer

Drone pilots may also need to complete photo and video courses. In addition to flying drones, drone pilots often edit their own photographs and videos.

Learning about photography ▶

Learning about plants

People who want to become drone pilots need to think about the work they wish to do with drones. Usually, a person who wants to work as a drone pilot will study two fields. For example, drone pilots who work on farms need to study farming or science in addition to their normal drone courses.

Drone pilots also have to complete continuing education courses. Technology is always improving and changing. Drone pilots have to grow and change with the technology they use.

◀ Working with a new kind of drone

Learning new technology

Drones at work

In the future, there will be even more job opportunities for drone pilots. Do you think you would like to become a drone pilot? It might be the perfect job for you!

Comprehension Questions

1. How can drones fly themselves?
    (a) With remote controls
    (b) By aircrafts
    (c) Using computer programs
    (d) Through videos

2. What can drones do?
    (a) Take photos
    (b) Take videos
    (c) Deliver packages
    (d) All of the above

3. Drone pilots can . . .
    (a) work as police officers.
    (b) monitor the building of skyscrapers.
    (c) fly regular aircrafts.
    (d) swim well.

4. Drone pilots can help protect wild animals from . . .
    (a) getting hurt.
    (b) being hungry.
    (c) getting cut down.
    (d) growing crops.

5. In addition to studying about drones, what should someone study to become a drone pilot on a farm?
    (a) Science
    (b) Building
    (c) Archaeology
    (d) Both a and c

Glossary

- **aerial** (adj.) in or from the air

- **aircraft** (n.) any vehicle that can fly and carry goods or passengers

- **archaeologist** (n.) a person who studies objects belonging to ancient times

- **coding** (n.) a system of computer programming instructions

- **edit** (v.) to prepare a photograph, movie, book, etc. by deciding what to include and by removing mistakes

- **lifeguard** (n.) a person who works at a beach or a swimming pool to rescue people who are in danger in the water

- **monitor** (v.) to watch and check something carefully

- **natural disaster** (n. phr.) a sudden and violent event in nature (such as an earthquake or flood) that causes a lot of damage

- **package** (n.) a box or bag sent by mail, carried easily, or given as a present

- **remote control** (n. phr.) something that allows you to operate a television, etc. from a distance

- **rescue** (v.) to save someone or something from a dangerous or harmful situation

- **rural** (adj.) relating to the countryside and not to towns or cities

Notes

Here are some interesting future jobs for drone pilots. Readers may enjoy researching these topics to learn more about the future in this field.

Phone companies are looking into ways of using drones in the place of cell phone towers. These drones could be used to restore phone service after natural disasters. Drone pilots will be needed to fly and program these drones.

Food and package deliveries may all be made by drones in the future. Someday, the pizza you order for dinner may be sent to your house by a drone pilot.

Oil companies are figuring out ways for drones to help detect gas leaks in pipelines. There are millions of miles of pipelines, and the drones could register the location of leaks. They could also share information with repair and cleanup crews.

Taxi companies may replace cars with drones that drive on streets instead of flying in the air. One day, we may call for drone taxis to take us from place to place.

In the future, we may see as many drones flying around as we now see cars driving around. We will need **drone traffic controllers**—a new job.

List of Books

LEVEL 1

1. Robotics Engineers
2. Cyber Security Experts
3. Medical Scientists
4. Social Media Managers
5. Asset Managers

LEVEL 2

1. Drone Pilots
2. App Developers
3. Wearable Technology Creators
4. Computer Intelligence Engineers
5. Digital Modelers

LEVEL 3

1. IoT Marketing Specialists
2. Space Pilots
3. Water Harvesters
4. Genetic Counselors
5. Data Miners

LEVEL 4

1. Database Administrators
2. Nanotechnology Research Scientists
3. Quantum Computer Scientists
4. Agricultural Engineers
5. Intellectual Property Lawyers

"The future of the economy is in STEM. That's where the jobs of tomorrow will be."

James Brown (Executive Director of the STEM Education Coalition in Washington, D.C.)

Data from the US Bureau of Labor Statistics (BLS) support that assertion. Employment in occupations related to STEM—science, technology, engineering, and mathematics—is projected to grow to more than 9 million by 2022 [in the US alone] . . . Overall, STEM occupations are projected to grow faster than the average for all occupations.

from *STEM 101: Intro to Tomorrow's Jobs* US Bureau of Labor Statistics